Zero Acceptance Number Sampling Plans

Fourth Edition

Nicholas L. Squeglia

ASQ Quality Press
Milwaukee, Wisconsin

Zero Acceptance Number Sampling Plans
Fourth Edition

Library of Congress Cataloging-in-Publication Data
Squeglia, Nicholas L.
    Zero acceptance number sampling plans / Nicholas L. Squeglia.—
4th ed.
      p. cm.
    Third ed. pub.: Milwaukee: American Society for Quality Control,
1986 under title: Zero acceptance number C=0 sampling plans.
    ISBN 0-87389-305-0
    1. Acceptance sampling.   I. Squeglia, Nicholas L. Zero
acceptance number C=0 sampling plans.   II. Title.
TS156.4.S68   1994
658.5'62—dc20   93-46817
                                                                     CIP

© 1994 by ASQ
All rights reserved. No part of this book may be reproduced in any form or by any means, electronic, mechanical, photocopying, recording, or otherwise, without the prior written permission of the publisher.

10 9 8 7

ISBN 0-87389-305-0

Acquisitions Editor: Susan Westergard
Production Editor: Annette Wall
Marketing Administrator: Mark Olson
Cover design by Linda Shepherd.
Printed and bound by BookCrafters, Inc.

ASQ Mission: To facilitate continuous improvement and increase customer satisfaction by identifying, communicating, and promoting the use of quality principles, concepts, and technologies; and thereby be recognized throughout the world as the leading authority on, and champion for, quality.

For a free copy of the ASQ Quality Press Publications Catalog, including ASQ membership information, call 800-248-1946.

Printed in the United States of America

 Printed on acid-free recycled paper

American Society for Quality
ASQ

Quality Press
611 East Wisconsin Avenue
P.O. Box 3005
Milwaukee, Wisconsin 53201-3005
414-272-8575
Fax 414-272-1734
800-248-1946
Web site http://www.asq.org

CONTENTS

Preface v

Introduction 1

Attribute Sampling Plans 2

Nonstatistical Sampling Plans 2

Relationship of C=0 Plans to MIL-STD-105E Plans 3

    Estimating Potential Savings 7

    Why Constant Sample Sizes Are Not Used 7

    Summary for Use 8

Use of the C=0 Plans Table 8

    Physically Taking the Sample 8

C=0 Sampling Plans Table 9

Comments of the A.O.Q.L. 10

    Table of A.O.Q.L. Values 11

    A.O.Q.L. Comparisons 12

Background Information 13

    Acceptance by Government Agencies

    MIL-Q-9858A and MIL-H-50

Adjustments from MIL-STD-105E 16

Sampling Plan "Switching" Comments 16

Operating Characteristic Curves & Values 17

Small Lot Size Supplement 33

Derivation 33

Small Lot Size Table 34

PREFACE

For many years the acceptable quality level (A.Q.L.) concept was used largely because of the influence of MIL-STD-105 and its revisions. With the immense worldwide competition, greater demands by customers, and the fact that quality control does not cost, it pays, the thinking has realistically changed to zero defects, and A.Q.L.'s are no longer the rule but the exception. Many companies are striving for zero defects through statistical process control, improved processes, closed loop inspection systems, and other means. In today's environment the A.Q.L. concept does not appear compatible.

Sampling plans with zero acceptance numbers are continually becoming more popular. The sampling plans in this book (C=0) actually represent a revision in 1963 of similar plans developed by the author in 1961. Because of the widespread use of MIL-STD105C in 1961, it seemed the only way to depart from this standard would be to develop a set of plans that could favorably be compared with the military standard. The C=0 plans were developed and implemented in a medium-size plant that did both military and commercial work. Although the plans were not formally approved, there was no opposition to them.

In 1963, MIL-STD-105D appeared and the C=0 plans were updated and revised. This time, the plans were introduced to a large aerospace manufacturer with a staff of resident government quality control representatives. It was necessary to put on a formal presentation and explain the C=0 plans in great detail. The military agreed to accept the plans on a trial basis. While the plans were targeted essentially to the lot tolerance percent defective (as stated in the LQ table) of the military standard, there were departures from these targets in several instances. These special adjustments were necessary to maintain the logic of the C=0 plans. These adjustments were highlighted during the presentation to the military with color slides.

The results of the trial period in this company were excellent. Not only were the savings significant, but there was a significant reduction in problems in assembly. A check with the company in 1983 (20 years later) revealed the C=0 plans were still being used.

The C=0 sampling plans were first presented in a national publication in 1965 (N. L. Squeglia, "Sampling Plans for Zero Defects." *Quality Assurance* 4, 28 [August 1965]). Because of the inquiries and interest from the article, a book was published in 1969 that described the plans in more detail and contained operating characteristic (O.C.) curves. The continual interest in the C=0 plans and primarily encouragement from Professor N. L. Enrick of Kent State University resulted in the publication of the second edition in 1981. The C=0 plans remained unchanged. As with this edition, the second edition was published to provide more information.

During 1983, a survey was conducted to get some idea of the extent of savings by users of the C=0 plans who switched from MIL-STD-105D. A few said it was too early to tell, but the majority reported a range of savings from 8% to 30%, with an average of 18%. Of course, the extent of savings is based on the lot sizes and index value (associated A.Q.L.) used. The larger the lot and index value, the greater the savings. It is not necessary to implement the plans to determine the savings. The savings potential can be evaluated from past data as later described.

While the hypergeometric distribution was used originally to maximize mathematical accuracy, it is the author's humble opinion that the most important feature of the plans is the philosophy of zero defects.

The C=0 sampling plans are in wide use throughout the United States. In 1983, the C=0 plans became a part of the Department of Defense's DLAM 8200.2 for use by government DCAS quality assurance representatives. In 1989, MIL-STD-105E superseded 105D. This revision placed emphasis on the use of 105E as a guide in developing inspection strategies and also recognized the limitation of the A.Q.L. concept. The sampling plans have not been changed. This fourth edition of the C=0 sampling plans updates the references to the 105 revisions. Also, a small lot size table was added.

Nicholas L. Squeglia

ZERO ACCEPTANCE NUMBER SAMPLING PLANS

INTRODUCTION

The zero acceptance number plans developed by the author were originally designed and used to provide equal or greater consumer protection with less inspection that the corresponding MIL-STD-105 sampling plans. In addition to the economic advantages, these plans are simple to use and administer. Because of these advantages, and where greater emphasis is being placed on zero defects and product liability prevention, these plans have found their place in many commercial industries, although originally developed for military products. Essentially, these commercial applications have resulted in the plans "standing on their own," and are therefore also used extensively where there is no connection to the MIL-STD-105E plans.

The derivation of these plans is covered in detail. It is important, however, to reemphasize that although the derivation involves considerable comparison with MIL-STD-105E, these C=0 plans are not limited to applications involving industries that are presently using the military plans.

There is no specific sampling plan or procedure that can be considered best suited for all applications. It is impractical to cite all of the applications where these C=0 plans are used. Some of these are machined, formed, cast, powered metal, plastic, and stamped parts; and electrical, electronic, and mechanical components. They have found application in receiving inspection, in-process inspection, and final inspection in many industries. **Wherever lot-by-lot sampling potential exists, regardless of product, the C=0 plans may be applicable.**

Quite often, the basic objective of sampling is overlooked. The primary objective is derived from the question, "why sample?" Most of us are aware of the fact that sampling is simply employed to provide a degree of quality protection against accepting nonconforming material. Further, we also know that what we are continually striving for is 100% good product. Assuming our inspection capability is 100% efficient in detecting nonconformances, the only way to assure 100% good product is to 100% inspect everything. This, then is the objective of sampling: **We sample because it is impractical in most cases to perform 100% inspection.** What we are seeking, therefore, are sampling plans that economically provide us with a reasonable amount of protection to ensure 100% good quality. There are times when something less that 100% good product is considered acceptable; in other words, there are times when we **knowingly** accept defective product. Such cases, however, should be treated on an exception basis.

This book provides a set of attribute plans for lot-by-lot inspection. The acceptance number in all cases is zero. This means that for some level of protection you select a certain size sample, as later described, and **withhold the lot** if the sample contains one or more nonconforming pieces.

The terminology "withhold the lot" is significant in that it does not necessarily mean rejection. Under these plans, the inspector does not automatically accept or reject the lot if one or more nonconformances are found. The inspector only accepts the lot if zero nonconformances are found in the sample. Withholding the lot forces a review and disposition by engineering/management personnel in regard to the extent and seriousness of the nonconformance. From this point, we will use the words "defective" and "defect" to describe nonconformances, regardless of whether the defective or defect is or is not fit for use. The word "defective" is commonly used in quality control to describe a part, component, item, or any other unit of product that contains one or more defects. The word "defect" is commonly used to describe a particular nonconforming characteristic on a unit of product. For example, a particular item contains a slot of certain width and length, and also contains a hole. All of these characteristics may conform to specifications except the diameter of the hole. This nonconforming diameter is a defect, and the item is therefore defective.

ATTRIBUTE SAMPLING PLANS

The following information provides a brief overview of lot-by-lot attribute plans while relating them to other plans.

Two broad categories of sampling are (1) continuous and (2) lot by lot.

Continuous

Continuous sampling is often used when units of product are submitted one at a time. This can apply, for example, to assembly lines. Product moving along a conveyor can also be thought of as being a candidate for continuous sampling. Consider, for example, a functional test. Units are moved along some line and assembly progresses to completion at the end of the line. Upon completion, we have a functional test. The continuous sampling plan may call for a frequency check of, say, one unit out of five. Should the products be good, this frequency check will remain in existence. If, however, a unit is bad, 100% inspection is reverted to MIL-STD-1235A provides continuous sampling plans. They basically start off with 100% inspection. When we find a "bad" unit on the frequency inspection, 100% inspection is reverted to until we get the specified number of consecutive good results. At that point in time we again go on the frequency inspection, and so on.

Lot by Lot—Attributes

Lot by lot involves units of products that are presented in a group, batch, or lot for inspection, as opposed to being presented one at a time.

In these cases, a sample of a specified quantity is drawn and compared with some acceptance criteria, for instance, the quantity of defectives allowed in the sample by the sampling plan. MIL-STD-105E allows to a great extent defectives in the sample. For example, a particular sampling plan may call for a sample size of 125 and an acceptance number 2. If 2 or less defectives are detected in the sample of 125 pieces, the lot is accepted. If more that 2 defectives are detected in the 125-piece sample, the lot is rejected. MIL-STD-105E, like the C=0 plans, contains attribute sampling plans. Measurements of characteristics are not required.

The characteristics evaluated either conform or do not conform. Go/no go type gages are often used in attribute plans.

Lot by Lot—Variables

Another lot-by-lot sampling procedure involves the analysis of measured characteristics MIL-STD-414 (ANSI/ASQC Z1.9) is a variables type sampling plan procedure. Variables sampling compared with attribute sampling essentially involves the inspection of a smaller sample size to obtain the same protection afforded by an attribute plan. The economics of the smaller sample size, however, are quite often offset by the calculating involved and the need for obtaining and recording measurement. Where variables data is required form an inspection operation, however, variables plans should definitely be considered. It should be noted that MIL-STD-414 plans assume normality in all cases, and that if a lot is rejected on the basis of calculation but no defectives are detected in the sample the inspector is required to use the corresponding MIL-STD-105E attribute sampling plans.

NONSTATISTICAL SAMPLING PLANS

There are cases where we can virtually assure zero defects although the sample size cannot logically be defined in terms of statistical risks. Such sample sizes are generally exceptionally low for the more important characteristics and, therefore, a knowledge of the process control features is required.

In order to avoid any confusion in justifying such sample sizes on inspection plans, specific notations should be used to avoid any tie-in with statistical risks. The reason for such a selection should be noted, either directly on the plan or in the quality engineering standards.

An example might be a stamping operation where just the first and last piece from a lot are inspected dimensionally. A statistical sample may be taken for visual inspection only.

The higher index values in the C=0 plans are also used where favorable process control has been demonstrated and just an audit is required. Although the statistical risks seem high, the risks from a practical standpoint would be exceptionally low.

RELATIONSHIP OF C=0 PLANS TO MIL-STD-105E PLANS

The MIL-STD-105E sampling plans are A.Q.L. oriented. Essentially the A.Q.L. (acceptable quality level) is a specified percent that is considered to be good quality. In any sampling plan an operating characteristic curve can be generated to define the risks of accepting lots with varying degrees of percent defective.

When we use the A.Q.L. concept, such as in the MIL-STD-105E plans, we have a high probability of acceptance associated with this A.Q.L. percentage. Normally, this is in the order of a .90 to .98 probability of acceptance level. The risk of rejecting this A.Q.L. percentage is in the order of 0.10 to 0.02 probability level. This rejection risk is called the producer's risk.

The assumption in employing the A.Q.L. concept is that some agreement has been reached between the producer and the consumer. And, because sampling is used, the producer must assume a risk of having a lot rejected although the actual percent defective in the lot is equal to or less than specified A.Q.L.

If no prior A.Q.L. agreement exists, and sampling is to be performed simply because 100% inspection is impractical, then overinspection is usually the result when the MIL-STD-105E plans are used. **Also, when we sample inspect because 100% is impractical, we would expect to inspect a smaller number of pieces on less critical characteristics.**

To illustrate this point, consider the following example. Let us assume we are presently using the MIL-STD-105E plans. A 1.0%t A.Q.L. is used for major characteristics, and a 4.0% A.Q.L. is used for minor characteristics, although no consumer/producer A.Q.L. agreement exists. Let the lot size equal 1300. From MIL-STD-105E based on the lot size, we must take a sample size of 125 pieces, with 3 defectives allowed for the 1.0%t A.Q.L. and 10 defectives allowed for the 4.0% A.Q.L.

FROM MIL-STD-105E

A.Q.L.	SAMPLE SIZE	ACC. NO.
1.0%	125	3
4.0%	125	10

We see that for a less critical characteristic the sample size remains the same. The difference is in the acceptance number. That is, for the 1.0% A.Q.L. we can accept the lot if the sample contains 3 or less defectives. For the minor characteristic we can accept the lot if the sample contains 10 or less defects.

Because the sample size is the same, **there is no reduction in pieces inspected on the minor characteristics.**

Now, there is another part of the O.C. curve for a sampling plan which is often overlooked. (A sampling plan consists of a sample size and acceptance criteria.) An O.C. curve for a sample size of 125, acceptance number = 10 is shown as Figure 1.1. The A.Q.L. and producer's risk, as previously described, are shown. Also, there is a parameter called the lot tolerance percent defective (L.T.P.D.). This L.T.P.D. is considered poor quality. Several sampling plans can have O.C. curves passing through the same A.Q.L./producer's risk point. For each of these plans, however, there will be a different L.T.P.D. at some constant probability of acceptance level. This probability of acceptance level corresponding to the L.T.P.D. is usually low, with 0.10 being widely accepted. This probability level is called the consumer's risk.

The user of the sampling plan, therefore, must select a plan that will provide reasonably good protection against accepting lots with a percent defective not too much greater that the A.Q.L.

With the A.Q.L./producer's risk point fixed, the closer the L.T.P.D. gets to the A.Q.L., the larger the sample size and acceptance number become. This is illustrated in Figure 1.2.

If we are satisfied, however, with the specified L.T.P.D. of MIL-STD-105E, and are not concerned with the A.Q.L. concept, we can get equal or greater protection than the MIL-STD-105E plan by using corresponding C=0 plans.

Figure 1.2 shows the effects of acceptance numbers on O.C. curves. With the acceptance number set to zero, we have greater protection at the L.T.P.D. level with a sample size of 18, as compared with a sampling plan from MIL-STD-105E that has a sample size of 125 with an acceptance number of 10.

Now, let us compare a set of C=0 plans from Table 1-a with the previous MIL-STD-105E example used.

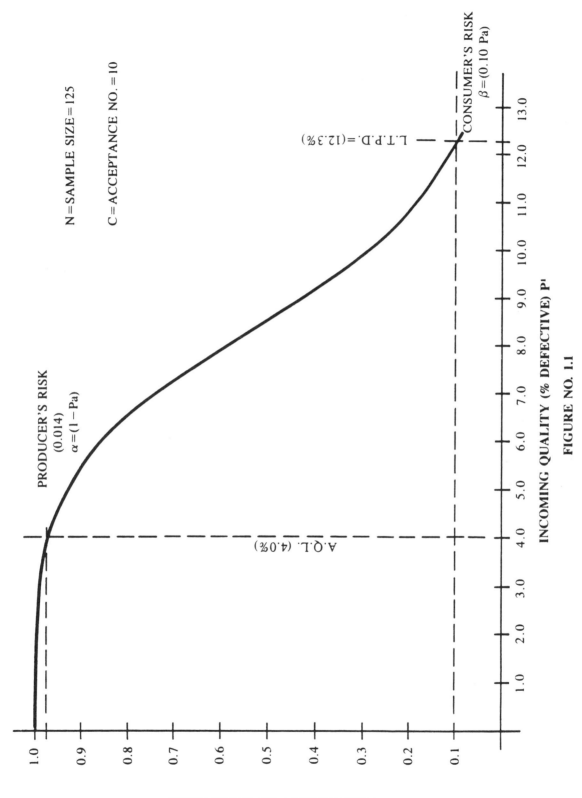

FIGURE NO. 1.1

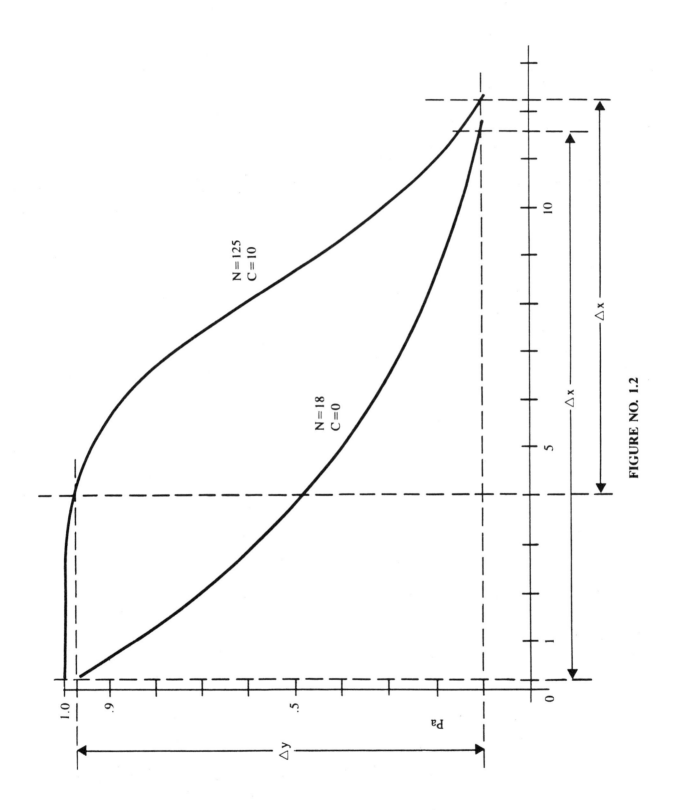

FIGURE NO. 1.2

MIL-STD-105E

	A.Q.L.	SAMPLE SIZE	ACC. NO.
	1.0%	125	3
	4.0%	125	10

C=0 PLAN

	ASSOC. A.Q.L	SAMPLE SIZE	ACC. NO.
	1.0%	42	0
	4.0%	18	0

The C=0 plans provide equal or greater L.T.P.D. protection at the 0.10 "consumer's risk" level. We also see that less inspection is performed on less critical characteristics.

All of the C=0 plans in the table are **"associated"** with the A.Q.L.'s of MIL-STD-105E. In all of these plans equal or greater protection is afforded to the consumer than the MIL-STD-105E plans. This method of developing the plans provides for simple conversion from the MIL-STD-105E plans to the C=0 plans. The table labels these associated A.Q.L.'s as "index values," because, of course, they are not A.Q.L.s.

It should be pointed out that the question of rejecting more lots under these plans may arise, because of the zero acceptance numbers. Aside from experience which has shown us that in fact considerable savings can be derived from using these plans, consider the following:

If your quality is very bad, acceptance numbers greater than zero will not be much help.

When you allow acceptance numbers greater than zero in your plan, you are in effect authorizing an inspector to accept parts which may not be usable. The zero acceptance number forces a review of any defectives by qualified personnel in order for a proper disposition to take place.

If you are striving for zero defects, how can you allow defectives in your sampling plans?

ESTIMATING POTENTIAL SAVINGS

In your particular situation, you can obtain an estimate of the cost reduction from switching from the MIL-STD-105E plans to the C=0 plans by examining your past records. Compare these records with the C=0 plans. For example, if MIL-STD-105E called for a sample size of 125, C=3, and the records show you inspected 125 pieces and found no defectives, then your comparison is the 125 pieces against the 42 pieces in the C=0 plans. Also of course, if you found 4 bad pieces in the 125 pieces sample, the chances are pretty high that one of these 4 were detected in the first 42 inspected. In other words, the lot would have been rejected with fewer pieces inspected.

Some actual inspection results are shown and discussed under "Background Information."

WHY CONSTANT SAMPLE SIZES ARE NOT USED

As discussed in this book, the C=0 plans were designed essentially to be equal or greater in Consumer and Average Outgoing Quality Limit protection than the corresponding MIL-STD-105E plans. The C=0 plans contain other features of the MIL-STD-105E plans. Within a particular "A.Q.L." column, the operating characteristic curves actually differ for the most part as the lot size increases. The reason for this common feature, in addition to satisfying the statistical relationship, is that it is generally considered more practical to obtain greater protection on larger lot sizes.

The use of constant sample sizes often results in a combination of over-inspection and under-inspection for broad lot size ranges.

IN SUMMARY, C=0 PLANS ARE USED WHEN:

1. Manufactured parts are expected to completely conform to specification requirements.
2. Less inspection is desired on less critical characteristics.
3. Sampling is performed because 100% inspection in general on all characteristics of all parts is impractical.
4. Inspectors, as a general rule, are not allowed to knowingly accept non-conforming products.
5. Interim inspections until a problem is corrected
6. Auditing is required for various reasons, i.e.:
   - Audit of stock for assurance
   - Audit of items for potential transit damage
   - Suppliers undergoing certification
7. Visual inspection in general

USE OF THE C=0 PLANS TABLE

Assume you are presently working with MIL-STD-105E plans and have established A.Q.L. levels.

A particular lot of 700 pieces is to be inspected. Normally, you would use, for example, a 1.0% A.Q.L. in accordance with MIL-STD-105E.

Under the C=0 plans, from the left hand column in table 1-a we find that 700 lies within the lot size range of 501-1200. Reading across, and down from the index value (associated A.Q.L.) of 1.0% we find our sample size to be 34.

From this lot of 700 pieces we take a sample size of 34, and if one or more defectives are found in the sample we withhold the lot.

If you are not presently working with MIL-STD-105E, then someone must designate the type of plans to use in terms of index value. This designation depends upon your particular situation, and requires product process knowledge. However, once you select the desired index value, the procedure for taking the sample is identical to that previously cited.

PHYSICALLY TAKING THE SAMPLE

A representative sample is necessary to assure reliable results. One of the most common ways to obtain a representative sample is by random sampling. Randomness is achieved only when each piece and combination of pieces in the lot has an equal chance of being selected for the sample. Such a sample may be drawn in several ways. When material is packaged in an orderly manner, or laid out on a bench in groups or rows, each unit may be numbered from 1 to the total in the lot, and the sample selected by use of a random number table. For bulk-packed material, a bench may be marked in numbered areas on which the units are evenly spread. The units for the sample can be selected by reference to the numbered zone in which they are located. Many variations of such approaches have been used with success.

If random sampling is not appropriate, common sense should prevail in obtaining a representative sample. Stratified sampling, for example, should be considered.

C=0 SAMPLING PLANS
INDEX VALUES
(ASSOCIATED AQLS)

LOT SIZE	.010	.015	.025	.040	.065	.10	.15	.25	.40	.65	1.0	1.5	2.5	4.0	6.5	10.0
								SAMPLE SIZE								
2 to 8	*	*	*	*	*	*	*	*	*	*	*	*	5	3	2	2
9 to 15	*	*	*	*	*	*	*	*	*	*	13	8	5	3	2	2
16 to 25	*	*	*	*	*	*	*	*	*	20	13	8	5	3	3	2
26 to 50	*	*	*	*	*	*	*	*	32	20	13	8	5	5	5	3
51 to 90	*	*	*	*	*	*	80	50	32	20	13	8	7	6	5	4
91 to 150	*	*	*	*	*	125	80	50	32	20	13	12	11	7	6	5
151 to 280	*	*	*	*	200	125	80	50	32	20	20	19	13	10	7	6
281 to 500	*	*	*	315	200	125	80	50	48	47	29	21	16	11	9	7
501 to 1200	*	800	500	315	200	125	80	75	73	47	34	27	19	15	11	8
1201 to 3200	1250	800	500	315	200	125	120	116	73	53	42	35	23	18	13	9
3201 to 10,000	1250	800	500	315	200	192	189	116	86	68	50	38	29	22	15	9
10,001 to 35,000	1250	800	500	315	300	294	189	135	108	77	60	46	35	29	15	9
35,001 to 150,000	1250	800	500	490	476	294	218	170	123	96	74	56	40	29	15	9
150,001 to 500,000	1250	800	750	715	476	345	270	200	156	119	90	64	40	29	15	9
500,001 and over	1250	1200	1112	715	556	435	303	244	189	143	102	64	40	29	15	9

*Indicates entire lot must be inspected

NOTE: The Acceptance Number in all cases is ZERO.

TABLE NO. 1-a

COMMENTS ON THE A.O.Q.L.

The average outgoing quality limit (A.O.Q.L.) is the maximum average outgoing quality from an inspection station for a particular sampling plan. For example, Figure 1.3 shows an A.O.Q. curve for the sampling plan N=50, C=0. The A.O.Q.L. is designated. **The A.O.Q.L. values for the C=0 plans are shown on Table 1-b. All of the A.O.Q.Ls for the C=0 are virtually the same or less than the corresponding MIL-STD-105E plans. In most cases the A.O.Q.L.s are lower, which of course is more desirable.**

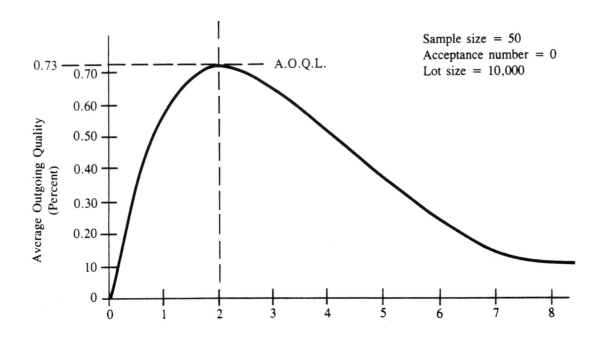

INCOMING QUALITY (PERCENT)
FIGURE NO. 1.3

INDEX VALUES
[ASSOCIATED A.Q.L.(%)]

LOT SIZE	.010	.015	.025	.040	.065	.10	.15	.25	.40	.65	1.0	1.5	2.5	4.0	6.5	10.0
								A.O.Q.L. (Acceptance number in all cases is zero)								
2 to 8													2.8	7.7	13.8	13.8
9 to 15											0.38	2.1	4.9	9.8	15.9	15.9
16 to 25										0.37	1.4	3.1	5.9	10.8	10.8	16.9
26 to 50									0.41	1.1	2.1	3.9	6.6	6.6	6.6	11.5
51 to 90							0.05	0.33	0.74	1.4	2.4	4.2	4.8	5.7	6.9	8.8
91 to 150					0.05	0.21	0.49	0.90	1.6	2.6	2.8	3.1	5.0	5.9	7.1	
151 to 280					0.05	0.16	0.33	0.60	1.0	1.7	1.7	1.8	2.7	3.5	5.1	6.0
281 to 500				0.04	0.11	0.22	0.39	0.66	0.69	0.71	1.2	1.7	2.2	3.3	4.0	5.2
501 to 1200			0.02	0.04	0.09	0.15	0.26	0.43	0.46	0.47	0.75	1.1	1.3	1.9	2.4	3.3	4.6
1201 to 3200	0.02	0.03	0.06	0.11	0.17	0.28	0.30	0.31	0.49	0.68	0.86	1.0	1.6	2.0	2.8	4.1
3201 to 10,000	0.03	0.04	0.07	0.11	0.18	0.19	0.19	0.31	0.42	0.54	0.73	0.96	1.3	1.7	2.4	4.1
10,001 to 35,000	0.03	0.04	0.07	0.12	0.12	0.12	0.19	0.27	0.34	0.48	0.61	0.80	1.1	1.3	2.5	4.1
35,001 to 150,000	0.03	0.05	0.07	0.07	0.08	0.12	0.17	0.22	0.30	0.38	0.50	0.66	0.92	1.3	2.5	4.1
150,001 to 500,000	0.03	0.05	0.05	0.05	0.08	0.11	0.14	0.18	0.24	0.31	0.41	0.57	0.92	1.3	2.5	4.1
500,001 and over	0.03	0.03	0.03	0.05	0.07	0.08	0.12	0.15	0.19	0.26	0.36	0.57	0.92	1.3	2.5	4.1

TABLE NO. 1-b

A.O.Q.L. COMPARISONS

For informational purposes, the graph shown as Table 1-c shows a comparison of A.O.Q.L. values for various sample sizes and acceptance numbers. Infinite lot sizes were used in constructing this graph, and, with the limitations of most graphs, the A.O.Q.L. values read approximate.

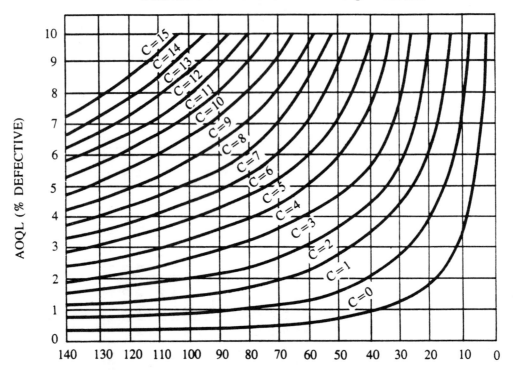

CURVES FOR DETERMINING AOQL VALUES

TABLE NO. 1-c

BACKGROUND INFORMATION

Acceptance by Government Agencies where MIL-Q-9858A is a requirement.

The C=0 plans have been accepted by Government Quality Control representatives for use in place of MIL-STD-105E.

Reference is made to the following which permits this authorization.

MIL-Q-9858A

Par. 6.6 STATISTICAL QUALITY CONTROL AND ANALYSIS

". . . The contractor may employ sampling inspection in accordance with applicable military standards and sampling plans (e.g., from MIL-STD-105, MIL-STD-414, or Handbooks H 106, 107 and 108). If the contractor uses other sampling plans, they shall be subject to review by the cognizant Government Representative. Any sampling plan used shall provide valid confidence and quality levels."

Note: There is no restriction or requirement to have "other" sampling plans match any specifics of MIL-STD-105E.

MIL-HANDBOOK-H-50
EVALUATION OF A CONTRACTOR'S QUALITY PROGRAM
(EXCERPT B. APPLICATION - PAGE 31)

"Some contractors find it advantageous to design their own sampling plans. Usually a qualified mathematician or statistician develops such plans to assure that they are valid and effective. The government must have such assurance; thus MIL-Q-9858A requires that the derivation, confidence level, protection and all other features of contractor-designed sampling be made known to the responsible Government authority upon request."

COMMENTS: This booklet provides sufficient information on the validity of these plans by means of the operating characteristic curves, A.O.Q.L. tables, and other information presented throughout this book.

Certain anomalies exist in MIL-STD-105E which are discussed and compensated for in the C=0 plans. The C=0 plans operating characteristic curves were calculated by a computer on the basis of the hypergeometric distribution. MIL-STD-105E utilized the binomial and Poisson distributions to calculate operating characteristic curves as an approximation of the hypergeometric distribution.

Some Actual Data Based Upon The Results of a Large Receiving Inspection Department Using the C=0 Plans.

During the 1960's a lengthy presentation was made to Government Representatives on the C=0 plans. Although previous studies had shown lower appraisal costs and lower assembly failure quality costs, we were asked to take one month's current actual inspection results from a large receiving inspection and compare results.

Table 1-d Shows the results of 1141 lots inspected. Essentially, of the 46 lots rejected, 45 would have been rejected anyway if the MIL-STD-105D plans had been used because the number of allowable defects was exceeded. (We compared the corresponding sample sizes and acceptance numbers.)

Table 1-e Shows the number of lots in which the sample size was the same as MIL-STD-105D, greater than 105D, and less than 105D. The reasons for the "greater than" is explained later in light of the anomalies detected in 105D.

Tables 1-f Continues the analysis in terms of pieces received. Although there were more pieces screened, there were lower total inspection costs.

Tables 1-g Shows the total number of characteristics inspected on the 1,250,752 pieces from the 1141 lots. It is here where the savings become more pronounced.

As the plans had been in use for some time prior to this study, a significant improvement in purchased material was realized. The C=0 plans, unlike plans with allowable defect numbers, do not allow or imply that producers may submit defective material, thus, eliminating or minimizing any misinterpretation about "A.Q.L.'s."

It was mentioned under "Use of the C=0 plans", that an inspector "withholds" a lot if one or more defectives are found in the sample. The word "withhold" is significant because the defective condition may not be cause for rejection and/or screening. The decision to reject, screen, or accept should be based on a review by competent authority, keeping in mind that we are referring to non-conformances which may or may not be fit for use.

INSPECTION RESULTS FROM LARGE RECEIVING INSPECTION DEPARTMENT FOR ONE MONTH PERIOD

NO. OF LOTS INSPECTED	NO. OF LOTS REJECTED	NO. OF LOTS IN WHICH NO. OF DEFECTIVES IN C=0 SAMPLE EXCEEDED ACCEPT NO. IN MIL-STD-105D	NO. OF LOTS IN WHICH NO. OF DEFECTIVES IN C=0 SAMPLE DID NOT EXCEED ACCEPT NO. IN MIL-STD-105D
1141	46	45	*1

*ONE DEFECTIVE IN C=0 SAMPLE SIZE 11. CORRESPONDING MIL-STD-105D N=50, c=3

TABLE 1-d

RECEIVING INSPECTION FIGURES FOR 1 MONTH

TOTAL NUMBER OF LOTS INSPECTED	NUMBER OF LOTS IN WHICH C=0 SAMPLE SIZE IS:		
	THE SAME AS MIL-STD-105D	GREATER THAN MIL-STD-105D	LESS THAN MIL-STD-105D
1141	430	30	681
	37.7%	2.6%	59.7%

TABLE NO. 1-e

TOTAL NUMBER OF PIECES RECEIVED	SAMPLE PLAN	TOTAL NUMBER OF PIECES INSPECTED	%	INSPECTION DUE TO SAMPLING	INSPECTION DUE TO SCREENING	%
1,250,762	C=0	50,032	4.0	30,672	19,360	1.5
	105-D	80,436	6.4	62,332	18,104	1.4

TABLE NO. 1-f

RECEIVING INSPECTION FIGURES FOR 1 MONTH

TOTAL NUMBER OF CHARACTERISTICS ON 1,250,762 PIECES RECEIVED	SAMPLE PLAN	TOTAL NUMBER OF CHARACTERISTICS INSPECTED	%	NUMBER OF CHARACTERISTICS INSPECTED DUE TO SAMPLING	%	NUMBER OF CHARACTERISTICS INSPECTED DUE TO SCREENING	%
24,907,674	C=0	304,122	1.2	282,866	1.1	21,256	0.1
	MIL-STD-105D	1,013,489	4.1	993,179	4.0	20,310	0.1

TABLE NO. 1-g

ADJUSTMENTS FROM MIL-STD-105E

There are 8 sampling plans in MIL-STD-105E that contain apparent anomalies, which are compensated for in the C=0 plans. These particular plans do not provide for the smooth progression required for "Less inspection on less critical characteristics." As such, the C=0 plans contain large samples in these instances. Also, of course, where the acceptance number is zero in the MIL-STD-105E plans, the sample size is the same in the C=0 plans.

Take for example code letter "M" in the MIL-STD-105E plans. Reading from more critical to less critical A.Q.L.'s we have the following sampling size.

M=1250, 800, 500, 315, 200, 500, 315, actually, both the 200 and following 500 are out of balance, but the 500 can be adjusted mathematically to conform to the logic of the C=0 plans. Thus for A.Q.L.'s .010 through .040, the sample sizes are the same in both the C=0 and MIL-STD-105E sampling plans. At .065 A.Q.L. when M=200 occurs, the C=0 sampling plan sample size is 300, which is greater than the MIL-STD-105E sample size. In 0.10 and up, the C=0 sample sizes are smaller.

The other anomalies in MIL-STD-105E occur at H-0.4 A.Q.L., J-0.25 A.Q.L., K-0.15, L-0.10, N-0.04, P-.025, and Q-.015.

There are 5 cases where an adjustment was made in the downward direction to provide for a smooth progression.

For example, take code letter "D" from MIL-STD-105E at the 4.0% A.Q.L. The sampling plan is N=13, C=1. An equivalent C=0 plan would require a sample size of 7. However, if 7 were used in place of 5 we would not have a smooth progression. Consideration was given to increasing the preceding sample size (i.e.: the approach used for the 8 cases previously described) but because of the higher A.Q.L.'s and closeness of the A.O.Q.L.'s, the apparent increase in risks was not considered of practical significance. If anyone should have a strong aversion to this the following can be specified: N=4 (code letter C at 6.5), N=7 (code letter D at 4.0), N=11 (code letter E at 2.5), N=17 (code letter F at 1.5), and N=28 (code letter G at 1.0). If this is done and the logic of the smooth progression is to exist, then the preceding samples should be increased.

SAMPLING PLAN "SWITCHING" COMMENTS

Considerable thought was given to "tightened" and "reduced" inspection. A table of C=0 plans "tightened" inspection was calculated and prepared, but never made a part of the C=0 plans.

Actually, different operating charateristic curves are involved for tightened, normal and reduced. The reasoning for not "switching" in the C=0 plans was essentially three fold. Switching would have taken away from the simplicity of the plans. Most companies were using just the "normal" table and perhaps correctly so. Switching also assumes a continuing series of lots.

The original intent of going to "tightened" inspection is questionable in these days of quality programs. Tightened inspection was viewed as a means to force corrective action by increasing the chance of lot rejections.

On reduced inspection, for all practical purposes, the 105E plans are getting into "non-statistical" sampling plans. The main problem is that lots can be accepted that are much worse than the original intended plan.

OPERATING CHARACTERISTIC CURVES & VALUES

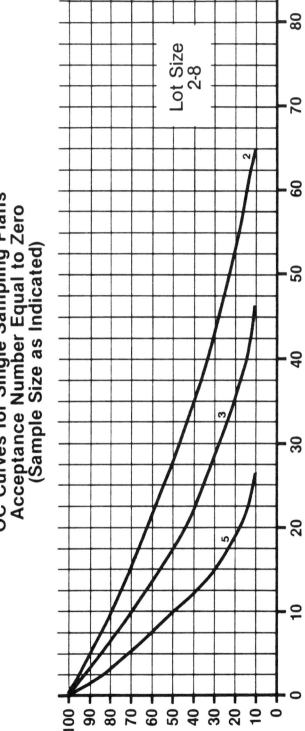

OC Curves for Single Sampling Plans
Acceptance Number Equal to Zero
(Sample Size as Indicated)

Lot Size 2-8

Sample Size	Probability of Acceptance						
	.10	.25	.50	.75	.90	.95	.99
2	63.7	46.9	27.5	12.5	5.00	2.50	0.50
3	46.7	32.5	18.3	8.33	3.33	1.67	0.33
5	26.0	18.3	10.0	5.00	2.00	1.00	0.20

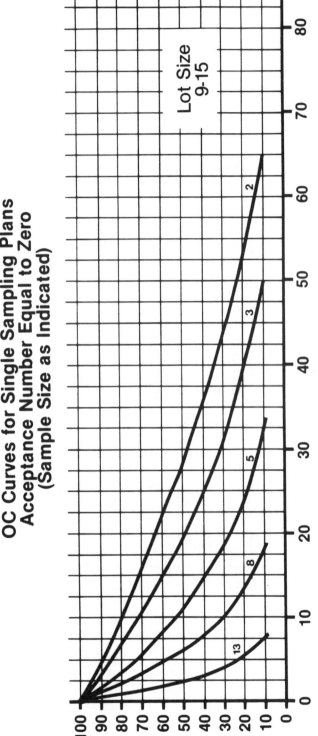

OC Curves for Single Sampling Plans Acceptance Number Equal to Zero (Sample Size as Indicated)

Lot Size 9-15

Sample Size	Probability of Acceptance						
	.10	.25	.50	.75	.90	.95	.99
2	66.0	48.3	28.3	12.9	5.00	2.50	0.50
3	50.0	34.5	19.2	8.61	3.33	1.67	0.33
5	31.8	20.8	11.3	5.00	2.00	1.00	0.20
8	18.7	12.1	6.25	3.13	1.25	0.62	0.12
13	8.46	5.77	3.85	1.92	0.76	0.38	0.07

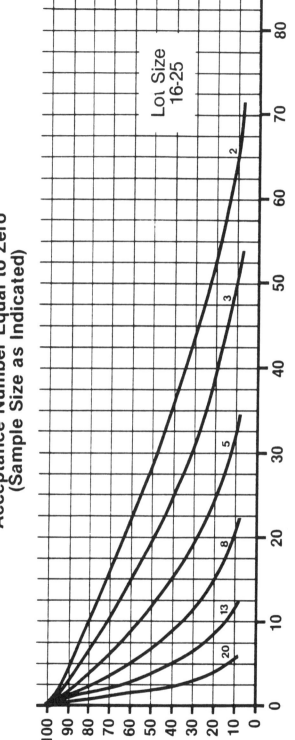

OC Curves for Single Sampling Plans
Acceptance Number Equal to Zero
(Sample Size as Indicated)

Lot Size 16-25

Sample Size	Probability of Acceptance						
	.10	.25	.50	.75	.90	.95	.99
2	67.0	49.0	28.7	13.1	5.04	2.50	0.50
3	51.4	35.5	19.8	8.80	3.33	1.67	0.33
5	33.9	22.3	11.9	5.20	2.00	1.00	0.20
8	21.4	13.7	7.18	3.13	1.25	0.62	0.12
13	11.9	7.54	3.85	1.92	0.76	0.38	0.07
20	6.40	3.75	2.50	1.25	0.50	0.25	0.05

OC Curves for Single Sampling Plans
Acceptance Number Equal to Zero
(Sample Size as Indicated)

Lot Size 26-50

Curves labeled: 3, 5, 8, 13, 20, 32

X-axis: Percent Defective (P')
Y-axis: Probability of Acceptance (Pa)

Sample Size	Probability of Acceptance						
	.10	.25	.50	.75	.90	.95	.99
3	52.5	36.3	20.2	8.97	3.39	1.67	0.33
5	35.4	23.2	12.4	5.38	2.00	1.00	0.20
8	23.2	14.8	7.72	3.31	1.25	0.62	0.12
13	14.2	8.91	4.59	1.92	0.76	0.38	0.07
20	8.73	5.42	2.82	1.25	0.50	0.25	0.05
32	4.60	2.94	1.56	0.78	0.31	0.15	0.03

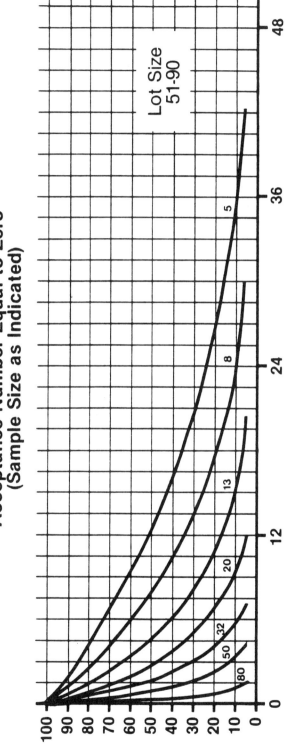

OC Curves for Single Sampling Plans Acceptance Number Equal to Zero (Sample Size as Indicated)

Lot Size 51-90

Sample Size	Probability of Acceptance						
	.10	.25	.50	.75	.90	.95	.99
5	36.1	23.7	12.7	5.47	2.04	1.00	0.19
8	24.0	15.3	7.98	3.39	1.26	0.62	0.12
13	15.1	9.44	4.86	2.05	0.76	0.38	0.07
20	9.70	5.99	3.06	1.29	0.49	0.25	0.50
32	5.68	3.48	1.80	0.78	0.31	0.15	0.03
50	3.17	1.98	1.00	0.50	0.20	0.10	0.02
80	1.23	0.93	0.62	0.31	0.12	0.06	0.01

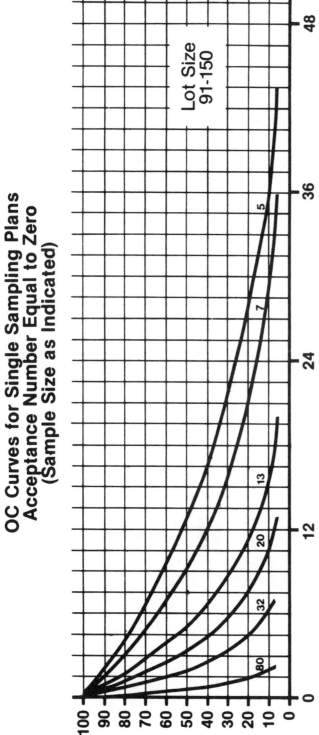

OC Curves for Single Sampling Plans
Acceptance Number Equal to Zero
(Sample Size as Indicated)

Lot Size 91-150

Sample Size	Probability of Acceptance						
	.10	.25	.50	.75	.90	.95	.99
5	36.4	23.9	12.8	5.52	2.06	1.01	0.19
7	27.5	17.6	9.24	3.95	1.47	0.71	0.14
13	15.6	9.71	4.99	2.10	0.77	0.38	0.07
20	10.2	6.27	3.19	1.34	0.49	0.24	0.05
32	6.21	3.80	1.92	0.81	0.31	0.15	0.03
80	2.0	1.24	0.62	0.31	0.12	0.06	0.01

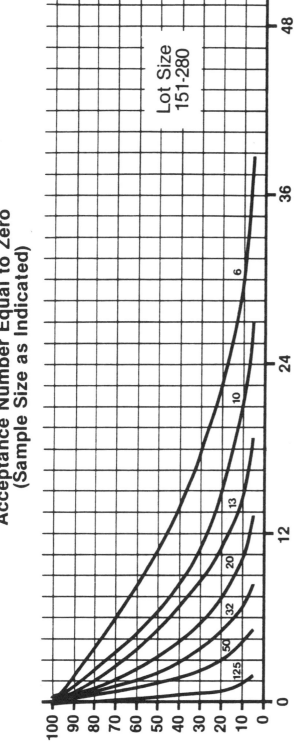

OC Curves for Single Sampling Plans
Acceptance Number Equal to Zero
(Sample Size as Indicated)

Lot Size 151-280

Sample Size	Probability of Acceptance							
	.10	.25	.50	.75	.90	.95	.99
6	31.6	20.4	10.8	4.64	1.72	0.84	0.16
10	20.2	12.7	6.59	2.79	1.03	0.50	0.09
13	15.9	9.90	5.08	2.14	0.79	0.38	0.07
20	10.5	6.47	3.29	1.38	0.51	0.24	0.05
32	6.55	4.00	2.03	0.84	0.31	0.15	0.03
50	4.10	2.49	1.26	0.53	0.19	0.09	0.02
125	1.39	0.85	0.43	0.20	0.08	0.04	0.00

OC Curves for Single Sampling Plans Acceptance Number Equal to Zero (Sample Size as Indicated)

Lot Size 281-500

Percent Defective (P')

Probability of Acceptance (Pa)

Sample Size	Probability of Acceptance						
	.10	.25	.50	.75	.90	.95	.99
7	27.8	17.8	9.36	4.00	1.48	0.72	0.14
11	18.7	11.7	6.04	2.55	0.94	0.46	0.09
16	13.2	8.17	4.17	1.76	0.64	0.31	0.06
29	7.41	4.54	2.30	0.95	0.35	0.17	0.03
50	4.28	2.60	1.31	0.54	0.20	0.10	0.02
125	1.59	0.96	0.48	0.20	0.08	0.04	0.00

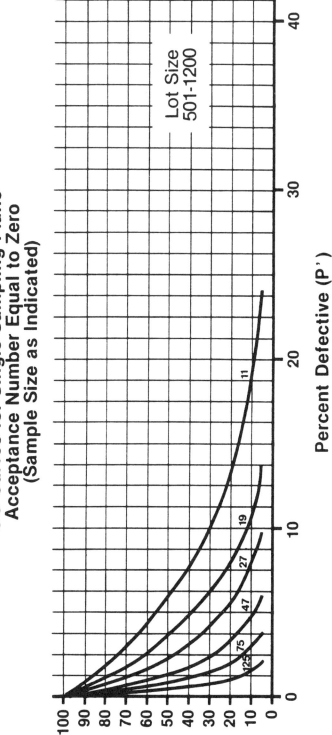

OC Curves for Single Sampling Plans Acceptance Number Equal to Zero (Sample Size as Indicated)

Lot Size 501-1200

Sample Size	Probability of Acceptance						
	.10	.25	.50	.75	.90	.95	.99
11	18.8	11.8	6.07	2.57	0.94	0.46	0.09
19	11.3	6.98	3.55	1.49	0.54	0.26	0.05
27	8.08	4.95	2.51	1.05	0.38	0.18	0.03
47	4.69	2.85	1.44	0.59	0.21	0.10	0.02
75	2.93	1.77	0.89	0.37	0.13	0.06	0.01
125	1.73	1.05	0.52	0.21	0.07	0.03	0.00

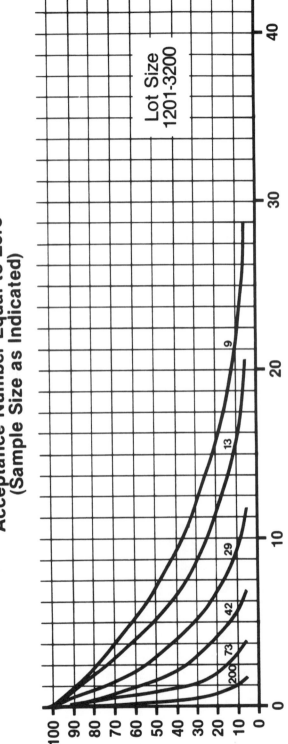

OC Curves for Single Sampling Plans
Acceptance Number Equal to Zero
(Sample Size as Indicated)

Lot Size 1201-3200

Sample Size	Probability of Acceptance						
	.10	.25	.50	.75	.90	.95	.99
9	22.4	14.2	7.37	3.13	1.16	0.56	0.11
13	16.1	10.1	5.17	2.18	0.80	0.39	0.07
23	9.46	5.82	2.95	1.24	0.45	0.22	0.04
42	5.29	3.22	1.62	0.67	0.24	0.12	0.02
73	3.07	1.86	0.93	0.38	0.14	0.06	0.01
200	1.11	0.66	0.33	0.13	0.05	0.02	0.00

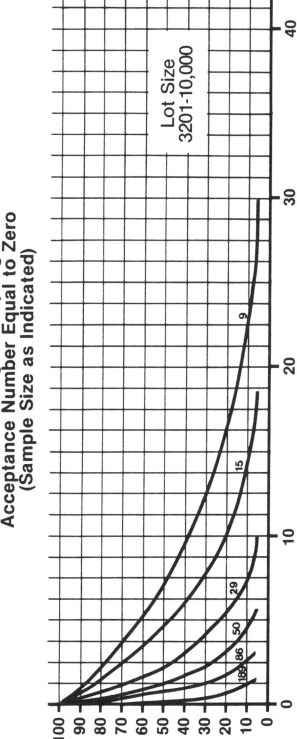

OC Curves for Single Sampling Plans
Acceptance Number Equal to Zero
(Sample Size as Indicated)

Lot Size 3201-10,000

Sample Size	Probability of Acceptance						
	.10	.25	.50	.75	.90	.95	.99
9	22.1	14.1	7.32	3.11	1.15	0.56	0.11
15	14.0	8.74	4.48	1.88	0.69	0.33	0.06
29	7.56	4.64	2.35	0.98	0.36	0.17	0.03
50	4.47	2.72	1.37	0.57	0.20	0.10	0.02
86	2.62	1.59	0.79	0.33	0.12	0.05	0.01
189	1.20	0.72	0.36	0.15	0.05	0.02	0.00

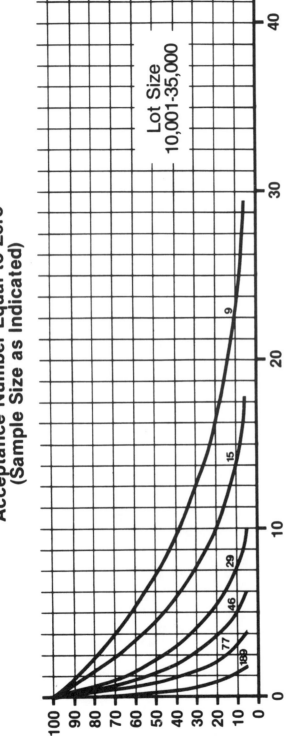

OC Curves for Single Sampling Plans
Acceptance Number Equal to Zero
(Sample Size as Indicated)

Lot Size 10,001-35,000

Sample Size	Probability of Acceptance						
	.10	.25	.50	.75	.90	.95	.99
9	25.6	15.4	7.38	3.02	1.14	0.55	0.10
15	15.4	8.54	4.40	1.85	0.68	0.33	0.06
29	7.61	4.58	2.33	0.97	0.35	0.17	0.03
46	4.80	2.93	1.48	0.61	0.22	0.11	0.02
77	2.91	1.77	0.89	0.37	0.13	0.06	0.01
189	1.20	0.72	0.36	0.15	0.05	0.02	0.00

OC Curves for Single Sampling Plans
Acceptance Number Equal to Zero
(Sample Size as Indicated)

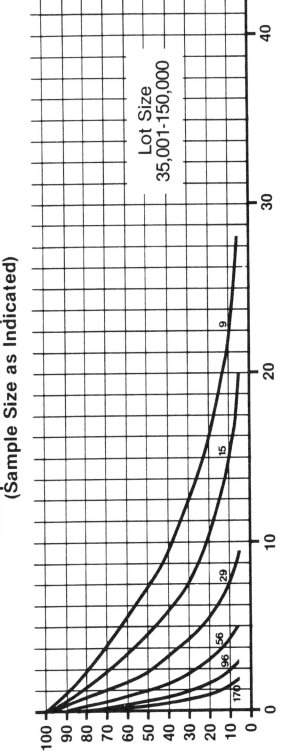

Lot Size 35,001-150,000

Sample Size	Probability of Acceptance						
	.10	.25	.50	.75	.90	.95	.99
9	25.6	15.4	7.70	3.20	1.14	0.55	0.10
15	15.4	9.24	4.62	1.92	0.69	0.33	0.06
29	7.94	4.78	2.39	0.98	0.36	0.17	0.03
56	4.11	2.48	1.24	0.51	0.18	0.09	0.01
96	2.40	1.44	0.71	0.29	0.10	0.05	0.01
170	1.35	0.81	0.40	0.16	0.06	0.03	0.00

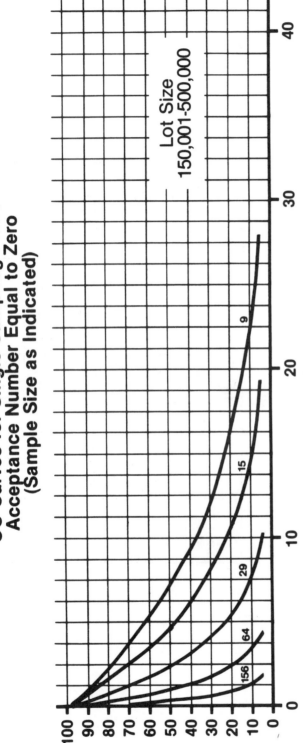

OC Curves for Single Sampling Plans Acceptance Number Equal to Zero (Sample Size as Indicated)

Lot Size 150,001-500,000

Sample Size	Probability of Acceptance						
	.10	.25	.50	.75	.90	.95	.99
9	25.6	15.4	7.70	3.20	1.08	0.52	0.10
15	15.4	9.24	4.62	1.92	0.66	0.32	0.06
29	7.94	4.78	2.39	0.99	0.35	0.17	0.03
64	3.60	2.17	1.07	0.44	0.10	0.07	0.01
156	1.48	0.88	0.44	0.18	0.06	0.03	0.00

SMALL LOT SUPPLEMENT

Many people have expressed a need for a special sampling plan table for small lots, when the associated A.Q.L. values they are using are 1.5 and below. Above 1.5, the main C=0 table works out well for small lot sizes. Any sampling plans developed for use with associated A.Q.L. less than .25 would (for the most part) not be valid. Therefore, in general, if you have broad lot size ranges and are using associated A.Q.L.'s in the .25 to 1.5 range, the small lot size sampling table should be of interest.

DERIVATION

The L.T.P.D. for the small lot sizes was targeted to the L.T.P.D. of the largest lot size range in which a constant sample size appeared. For example, (refer to the main C=0 table) for an associated A.Q.L. of 1.0, 13 was used at the 91-150 lot size range; for .65, 20 was used at the 151-280 range; and so forth.

In other words the sample sizes for the smaller lots shown in the supplement have essentially the same L.T.P.D. as the targeted L.T.P.D. As in the main C=0 table, the hypergeometric distribution was used.

The hypergeometric is a discrete distribution and necessitated interpolation to arrive at a Beta (consumer's risk) of essentially 0.10.

The small lot size supplement (associated A.Q.L.'s .25 thru 1.5) with lot sizes up to 35 do not give the same protection as MIL-STD-105E. However, it does give you the same protection as you would get on the targeted, or larger lot sizes.

Obviously, if you look at, for example, the lot size of 16-25 in the main table, you find that 100% inspection is required for the .40 associated A.Q.L. and therefore your protection is 100% as opposed to taking any sort of sample. Yet, the larger lot sizes involve samples. In the supplemental table, the sample size is 17. Thus, if you had a lot size of 25, you will inspect only 17, or, 32% fewer pieces, and get the same protection as if you were sampling in the lot size range of 151 to 280 of the main table C=0 plans or the MIL-STD-105E plans.

SMALL LOT SIZE SUPPLEMENT
ASSOCIATED A.Q.L.'S

LOT SIZE	.25	.4	.65	1.0	1.5
5-10	*	*	*	8	5
11-15	*	*	11	8	5
16-20	*	16	12	9	6
21-25	22	17	13	10	6
26-30	25	20	16	11	7
31-35	28	23	18	12	8

TABLE NO. 1-H